# Tiny House Builder

# How to Build a Simple Wooden House

# Step By Step Guide

# With Over 100 Pictures and Plans

I0484480

**Colvin Tonya Nyakundi and John Davidson**

*Prepping and Survival Series*

*JD-Biz Publishing*

Check out some of the other Entrepreneur Series books
Entrepreneur Series books on Amazon
Check out some of the Science of Living Series books
Science of Living Series on Amazon
Check out some of the Health Learning Series books
Health Learning Series on Amazon

# Table of Contents

# Introduction

It is everyone's dream to own a home in a quiet, secluded and serene environment. Owning such a home offers total privacy and a therapeutic experience that can't be found elsewhere. Regardless of whether it is the primary or secondary residence, it offers the perfect getaway during weekends and holidays. A simple and comfortable wooden house is what you need in order to have a feeling of serenity and privacy.

If you want to bond with your spouse, children or friends, there is no better way to do it than spend some time with them around a simple wooden house in a secluded environment. On top of bonding with family and friends, a wooden house provides the perfect resting place after a successful hunting trip.

Wooden houses can be built anywhere on earth regardless of the natural phenomena experienced there. For instance, you can build the house in areas prone to earthquakes and rest assured that nobody will be severely injured or die in case of an earthquake. Even if the house is brought to the ground by an earthquake, there is minimal probability of anybody being injured with wooden walls and roof.

Repairing wooden houses is also quite cheap when compared to repair of houses built with concrete, blocks, bricks or any other construction material. This means that you'll end up saving money by simply deciding to build a wooden house. It is also quite easy to alter the design of a wooden house than houses built using other materials. If you want special features in a home, all you need to do is make sure that you own a wooden house and then install all the features you want.

With the book "How to Build a Simple Wooden House," you'll have everything required to construct a wooden house anywhere in the world. The book contains step by step guidelines on how you can build such a house from scratch.

Start your journey to owning your dream home by reading the book: How to Build a Simple Wooden House!!!

## Required Tools and Construction Materials

Constructing a wooden house starts with the acquisition of the necessary tools and the right quantity of construction materials. Without the appropriate tool for each job, you'll end up wasting a lot of time and energy and the project will delay for no good reason. If you start the construction process with insufficient construction materials, the project may stall midway.

Timber is the most important raw material for anybody planning to construct a simple wooden house. Without this material, there is no way you can build a wooden structure even if you have all the other materials. You'll use the timber to construct walls, roof, and other support structures. Since timber is the core component in any wooden structure, you must be very cautious when purchasing it. This means that there are several factors that you need to consider before purchasing any piece of timber. Some of the factors you need to consider include the following:

- Durability of timber

Ask anybody who deals in timber and you'll be surprised to learn that timber from different trees is not necessarily the same. For instance, timber obtained from mahogany is not the same as that obtained from eucalyptus, blue gum or cypress trees. Some trees can withstand more external force without breaking while others develop cracks and eventually break with minimal force being applied on them. This means that you must consult the timber dealer so as to get advice on the most appropriate timber to use in constructing the house. Remember that different climatic and geographical conditions may require different types of timber. For instance, if you plan to construct a home in a wet area, you need timber that can withstand heavy rainfall without rotting or starting to leak water into the house. On the other hand, if you're thinking of building the house in an area that experiences strong winds, you must ensure that the selected timber can withstand the strong winds. It is therefore up to you to determine the climatic conditions in the area and then purchase the most appropriate type of timber.

- Is the timber treated?

Before the construction process commences, you need to survey the area and ensure there are no insects likely to destroy the timber. If there are insects in that area, you must ensure that the timber is treated with a chemical that repels the insects.

We used pressure treated wood for the floor joists because they would be next to the ground. We used railroad ties as a foundation for our small wooden house.

- Availability and cost of the timber

Regardless of the amount of cash in your bank account, you still need to save as much money as possible. This means that you need to consider the cost of timber in the area. Keep in mind that you cannot ignore the cost of timber as it is the main raw material and hence it will constitute the highest percentage of your total expenditure. You must therefore consider sampling several dealers and getting their average prices. This way you'll be able to note the most cost effective and convenient dealer in timber.

- How long has the timber been lying at the timber-yard?

The quality of newly harvested timber is not the same as the one that has been lying on the yard for several weeks or months. You must also consider the conditions under which the timber is stored by the dealer. For instance, timber that has been exposed to direct sunshine for several weeks is likely to be of low quality than that stored in a cool, covered area. You should always purchase newly harvested timber that is being stored in a cool, dry area, away from direct exposure to the sun's ultraviolet rays.

However, you don't need to worry about anything because timber is one of the raw materials that are readily available in most parts of the world. This means that you'll have a wide variety of options to choose from. You just need to ensure that you purchase timber in different sizes to fit different parts of the house.

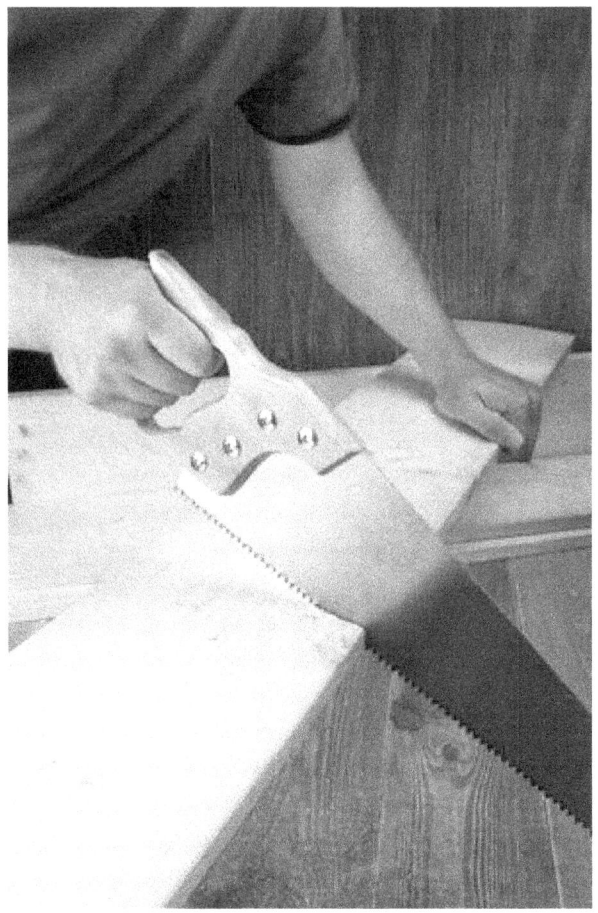

After acquiring timber, you need to purchase nails in different sizes. Nails are very important when constructing wooden structures as they're used in joining pieces of wood together. There are several types of nails, designed to be used on specific types of timber. For instance, it is quite difficult to join pieces of hard wood as they bend/crack due to the impact of the hummer. However, there are special types of nails that can be used to join two pieces of hard wood without bending or cracking them. This means that you should be guided by the type of timber when deciding the type of nail to buy. Depending on the type of timber, you can purchase the following types of nails:

- Round wire nail

The round wire nail is designed to be used for general purposes. However, you must be cautious when using this type of wood as it may split the timber and hence weaken your wooden house. Round wire nails are suitable for use when joining two pieces of soft wood or board.

- Oval wire nail

This type of nail can be used comfortably as it is designed to penetrate wood without splitting it. It is therefore the most ideal nail for use during joinery work involving hard wood.

- Lost head nail

The lost head nail is the most appropriate type of nail for places where you don't want the nails to be seen. It is designed with a head that can be hammered beneath the surface level and hence becoming unnoticeable.

Other types of nails that can be used in joining different types of wood include annular nail, sprig, hardboard pin, tack, panel pin and corrugated fastener.

Now that you have timber and nails, you still need iron sheets or roofing tiles, depending on what you want the roof to be made of. There are several types and brands of roofing structures in the market and hence it is up to you to choose the one that appeals to you the most.

When constructing wooden houses, you must ensure that you have a hand saw and a power saw. Remember that the timber needs to be cut into the appropriate sizes to fit in different parts as indicated in the house plan. The hand saw will be used in cutting small pieces of wood (small scale work) while the power saw will be used in large scale jobs such as when cutting posts and trusses.

There are several types of hammers designed to be used in different types of jobs. For optimal performance, you must ensure that you use the right type of hammer for the right job. Here are some of the hammers and where you can apply them:

- Claw hammer

The claw hammer is one of the most popular types of hammer and it is found in most households. This type of hammer is designed with a 'V' shaped cut-out for drawing nails from timber. When drawing nails using this hammer, you must be cautious because excessive force might end up weakening the joint.

- Ball peen hammer

This type of hammer is mainly used in closing rivets and shaping metals. It can be used when fitting gutters after the house has been constructed.

- Cross peen pin hammer

This is the most ideal type of hammer for light cabinet and joinery work in the construction process. For example, if you decide to fix wooden counter tops in the house, this will be the most appropriate type of hammer to use.

- Club hammer

Club hammer is a heavy duty type of hammer that is mainly used in light demolition work. Always keep in mind that you might need to demolish part of the wooden structure at some point in the construction process. You might therefore end up wasting a lot of your time if you don't have the club hammer. This hammer can also be used in driving steel chisels and masonry nails.

- Mallet

A mallet is made entirely of wood. It is used when tapping wood joints together so as to strengthen them. It can also be used in driving chisel.

There are also several other types of hammers designed for use in special occasions. These include brick, veneer, upholsterer, sprig and power hammers.

Putting down the plywood flooring

# Logistics of Building Small Wooden Houses

Building the walls

The management of workers and sourcing and transportation of raw materials can be quite challenging especially if you've never been involved in constructing wooden structures. However, you're guaranteed of a superb wooden house if you perfectly manage the logistics of building the house. All that you need is to know what to do, when to do it and how to do it.

Before starting to construct a house, you need a detailed building plan of the house you're about to construct. The building plan must clearly indicate the exact size of the house (number, orientation and location of rooms). The building plan should also indicate the different types of materials you intend to use in different parts of the house. The electrical and plumbing layout of the house should also be included in the building plan. A building plan helps you estimate the quantity of raw materials required to accomplish the task. It also helps in estimation of the money required to purchase raw materials and pay workers. If you have time constraints, a building plan will help you estimate the time it will take you to be through with the construction process.

See Appendix for complete small house plans.

Before you start the construction process, it is always advisable that you get approval from the necessary government agency. Without doing so, your house might be declared unsafe and hence earmarked for demolition. It is also possible that you might be fined or even imprisoned for violating the housing and urban development (HUD) code. Some of the factors that might influence the approval of your building plan include: the impact of the project on the environment, safety standards installed in the house and the HUD codes applicable in the area.

Now that your building plan has been approved by the government, all you need is enough cash to complete the project. You must never start a project if you don't have enough cash to complete it. If the money is not enough, you might want to consider applying for a bank loan or borrowing from friends and/or family. The money allocated to this project should be enough to purchase high quality raw materials, pay employees and finance finishing touches. It is also advisable that you have some extra cash stashed somewhere so that it can be used in case something comes up. For example, you might need to buy something that you forgot to include in the budget estimates. You might also be forced to repurchase broken parts such as roofing tiles.

During the construction of wooden houses, you don't have to buy all raw materials at the same time. It might be difficult to find a warehouse to store all the materials during the construction process. However, you can go ahead and buy all of them if you're guaranteed of a warehouse or store. If you decide to buy raw materials in bits, you must ensure that all the materials are readily available at the hardware store. Alternatively, you can organize with the store owner such that you purchase all the necessary materials and then pick them in bits. Always ensure that transportation is ready and efficient so that the construction doesn't stop due to lack of some or all of the construction materials.

The next phase involves hiring qualified, experienced and dedicated workers to help in the construction process. Keep in mind that the workers will do bulk of the work and hence they're the ones who'll determine how the house will look like once completed. This means that you must be very careful when selecting workers so as to ensure that you're working with people who can deliver exactly what you expect. If you don't have enough experience in building wooden houses, you might want to delegate the job of managing workers to a foreman. Before you begin the construction, it is always advisable that you consult other people who've been involved in construction before. This way you'll get to know what to expect and what not to expect as your house is being constructed.

With all construction materials, enough cash and personnel, you're ready to begin the process of constructing a simple wooden house. You can start by erecting poles in the corners of the house. The building plan should be able to guide you on where and how to place the posts. You'll then go ahead and fix the support structures between the poles. Before you put the walls, you can put trusses. Iron sheets, roof and walls are fixed after the poles and support structures have been installed in the house.

The next phase in the construction of wooden houses involves the installation of the plumbing and electrical system. A plumber will come in handy during the installation of pipes for distribution of water and gas in the house. The plumber must also install the sewerage and waste disposal system in the house. This process shouldn't be very difficult as the building schematics contain the layout of the plumbing system.

Installation of the electrical system is also one of the key areas during the construction of a simple wooden house. A faulty electrical system may result in frequent short circuits that may ignite a fire in the house. Remember that the house is made of wood, which burns easily and hence you must prevent fires by all means possible. This means that you must be cautious when hiring an electrician so as not to fall in the hands of inexperienced or unqualified people. After the plumbing and electrical systems have been installed, you need to install the flooring. The most appropriate type of flooring in these types of houses is wooden floors. The floor should be able to cover the pipes running from one point to another in the building.

With the electrical and plumbing system in place, you can go ahead and fix ceiling boards and interior features such as soft boards on each wall. The good thing with a wooden house is that you can alter the size and location of interior walls so as to fit your needs.

All that is remaining is for you to be connected to the national electricity grid. In case the national grid is located far, far away from the house, you can purchase a generator to supply the electrical power. Even if you're connected to the national electricity grid, you can still purchase a generator to act as a backup in case of a blackout.

The last step in construction involves finishing touches. These are special features installed in a house to make it look more appealing, make life in the house more comfortable and also to increase the value of the property. (Make sure that you read the chapter- in this book that - deals exclusively with the installation of finishing touches.)

# How to optimize your employees' performance

As you already know, workers play a critical role in the construction of small wooden houses. After you've delegated them duties, they'll be the ones to do the manual job. This means that they'll be the greatest determinants of the final structure. There are several things that you can do so as to ensure that your employees perform optimally.

To begin with, you must always give clear and precise instructions to each of your workers. This means that you have to notify them of what you expect them to do, how you expect them to do it and when you expect them to finish doing it. Without clear and precise instructions, employees might end up dragging the implementation of the project for no reason. It is always important that you give them reasonable timelines and expectations. They might end up doing a shoddy job if you overwork them.

In case of changes to the job description, you must notify the concerned employee(s) as soon as possible. For example, if you decide to alter the location of the kitchen, you need to notify the concerned worker(s) immediately so that they can adjust accordingly. This way you'll avoid demolitions and hence save on cash.

You also need to take care of your payment system if you want your employees to perform optimally. There is no way that a worker can do an excellent job if you constantly delay their pay or underpay them. Just ensure that you pay them at reasonable rates and the cash is available in time as agreed at the beginning of the construction. You can also occasionally give them bonuses as a way of boosting their morale and encouraging them to do their best.

After working for several hours, anybody needs a break so as to recoup the used energy. Manual workers are no exception to this rule and they're likely to perform better if they take a break after working for two-three continuous hours. They can take a 15 to 30 minute break while taking a snack or just resting. You also need to make sure that they go for lunch break for at least one hour. This way, they'll come back reenergized and willing to work even harder.

In this kinds of jobs (constructing wooden houses), workers are exposed to several dangers. For example, a worker might be injured by a claw hammer or a falling piece of wood. It is also possible that one of the workers might trip and fall injuring themselves. This means that you must always ensure that they're wearing protective clothing. Some of the necessary protective clothing includes a pair of safety boots, overalls, heavy duty gloves, dust masks, goggles and reflective jackets. It is also important to ensure that you have a fully equipped first aid kit within the vicinity. This might help save somebody's life before paramedics arrive.

So as to encourage the employees to work even harder and without fear, you can decide to cover them with insurance. This way an employee will be guaranteed of money to use in seeking medical attention if they're injured or fall sick while at work. In case of an accident that results in paralysis, incapacitation or even death, an employee will be guaranteed that their family will be taken care of. This means that they'll be willing to do anything to make sure that the project is completed successfully and without any delays.

When dealing with workers, it is always important to build on trust and give them space. This means that you shouldn't be monitoring all their movements so closely to the extent that they feel intimidated. Just make sure that they understand what you're expecting them to do and when you're expecting them to be through with the job. You can then leave them to do the job and come back for routine check-ins or when you've been called to clarify something.

When managing the construction of wooden houses, every employee must take responsibility for their actions. This means that there is no excuse for doing poor quality job or underperformance. Disciplinary action should be taken on any employee who's not performing as expected.

When several people are working together, it is possible that they might not agree on something. However, they should be able to solve their issues amicably. You must never condone rivalry between employees as it is likely to contribute to poor performance. In case of a disagreement between one or several employees, you should first try to mediate and solve the issue. However, if they can't agree to set aside their differences, you should lay off the one causing mayhem. The only way that the project can be concluded in time and to the best quality is if the employees co-exist in harmony.

# Necessary Finishing Touches

The outward appearance and attractiveness of a wooden house is highly influenced by the finishing touches applied on it. It might look as a simple job, but practically it is one of the most difficult and complex stages in building a house. Finishing touches require high levels of creativity for the house to stick out from all the other wooden houses. You might be forced to enlist the services of an experienced person if you don't have enough experience to install finishing touches on a house. The following are some of finishing touches that affect the appearance and state of the wooden house:

- Security features

Throughout the world, you can never be a hundred percent certain that a burglar won't try to break into your house. Rather than wait for somebody to come and still your valued possessions, you might as well install security features to keep thieves at bay. Security features should include an alarm system linked to emergency response units within the area. This way you can be sure that anybody who tries to break into your home will be caught before stealing anything.

- Smoke detectors

Unlike other construction materials wood can easily catch fire. Fires in wooden houses also tend to spread faster than in houses made of other materials such as concrete, bricks or blocks. This means that you must try to prevent a fire by all means. By installing smoke detectors, you can be able to detect a fire before it reaches uncontrollable levels. You should ensure that the smoke detectors are installed in such a way that emergency response systems are notified even if you're not at home.

- Solar panels and water heaters

Even though the initial installation cost might be high, a solar panel is quite cheap in the long run. A medium size solar panel can be able to power a standard size house (3 bedroom house) whenever there is enough sunshine. This means that you'll be able to significantly reduce the amount of cash you pay utility service providers. Rather than heat water using electricity, you can save huge amounts of cash by simply installing a solar water heating system. Currently, there are several companies offering solar panel and water heater installation services at reasonable rates. It is now up to you to identify one that will be able to install a system that satisfies your demand for power and hot water.

- Application of paint

Application of paint is also one of the finishing touches that should never be ignored. The type, quality and intensity of paint applied on the house will have a huge impact on how it looks like. This means that you must always be very careful when choosing the paint to apply in your newly build house.

- Cable television

Most people are always interested in knowing what is happening in their locality as well as in different parts of the world. You can get such information by simply watching the local TV station or tuning your radio. Apart from getting news, a TV can also be used for entertainment purposes among many other things. This means that you need to install cable television as part of the finishing touches in a newly built house.

- Wi-Fi router

The internet is something that has completely revolutionized the world and changed the way people do things. Right now, you don't have to waste your energy and time visiting libraries so as to get information about a given subject. All you need to do is search the internet and you'll find everything you need. The internet can be used for academic purposes, entertainment, doing business or even getting in touch with friends through social networking websites such as Facebook and Twitter. With these and many more advantages of being connected to the internet, there is no way you can ignore the importance of installing a Wi-Fi router in your home. This cheap, easy to install and readily available device will help you connect to the internet easily and conveniently.

- Borehole or well

A borehole or well can also be included as part of the finishing touches in a newly built house. They provide a constant source of water throughout the year. A borehole is even more important if the house is built far away from where you can be connected to piped water from utility service providers.

- Landscaping

Landscaping is the process of altering the natural and manmade features and plants so as to increase the esthetic value of the area surrounding the house. Landscaping is a very important part of the finishing touches as it determines the beauty in the entire area surrounding the house. Landscaping revolves around the type and nature of flowers, trees, herbs, vegetables, lawns and pavements around the home.

Finishing touches have a great impact on the value of the house. By simply installing the best and most outstanding finishing touches, you can significantly increase the value of your property and hence you're likely to make supernormal profits in case you decide to sell it. If you're thinking of using the wooden house as a lodge, you're likely to charge more cash when the finishing touches are splendid.

# Conclusion

When constructing a simple wooden house, it is always important to make sure that construction tools are well maintained and kept safely. Tools such as saws and hammers might end up hampering your ability to build splendid structures if they're not working efficiently. The only way you can maximize their efficiency is by making sure that they're always well maintained and kept where they can't be damaged. Proper storage will also help prevent injuries to anybody near them.

If you fully implement the tips on how to optimize your employees' performance, you can rest assured that the project will be completed in time and with minimal problems. You'll also get to save some cash if your workers are optimizing their potential.

After the house is complete, you must ensure that it is always well maintained. Maintenance involves replacement or repair of broken or damaged structures. The best thing about wooden houses is that maintenance is quite easy and you can do it yourself. As soon as any part is damaged, you should repair or replace it immediately. You must also nurture the habit of repainting the home as frequently as possible. This'll make it always look new and hence more valuable. Structures and features around the home should also be well maintained throughout the year. For example, you must ensure that all trees and flowers around the house are well trimmed and pruned.

Regardless of the safety features installed in a house, accidents can still happen. A fire can for example spread quickly before fire engines arrive to put it off. It is also possible that burglars can break into your house and get away with something precious even after you've installed security features. So as to ensure that fires and burglaries won't have a huge impact on your financial situation, you should cover the house with the appropriate insurance cover.

# Appendix

Download complete Blueprints at http://sdsplans.com

## 24 x 32 Small house Plans

Custom Cabin Design
#H274 24 x 32 Mountain Cabin
By Specialized Design Systems

Page 1    Cover Page
Page 2    Main Floor Plan
Page 3    Foundation Plan
Page 4    Elevation Plan
Page 5    Floor and Post Framing Plan
Page 6    Whole House Framing Section
Page 7    Cabinet & Stair Details
Page 8    Electrical Plans
Page 9    Second Floor Plan and Electrical
Page 10   Details

BUILDING CONTRACTOR HOME OWNER
TO REVIEW AND VERIFY ALL DIMENSIONS
SPECS AND CONNECTIONS BEFORE
CONSTRUCTION BEGINS. HOME TO BE
BUILT AS PER UBC LBC OR CURRENT CODE

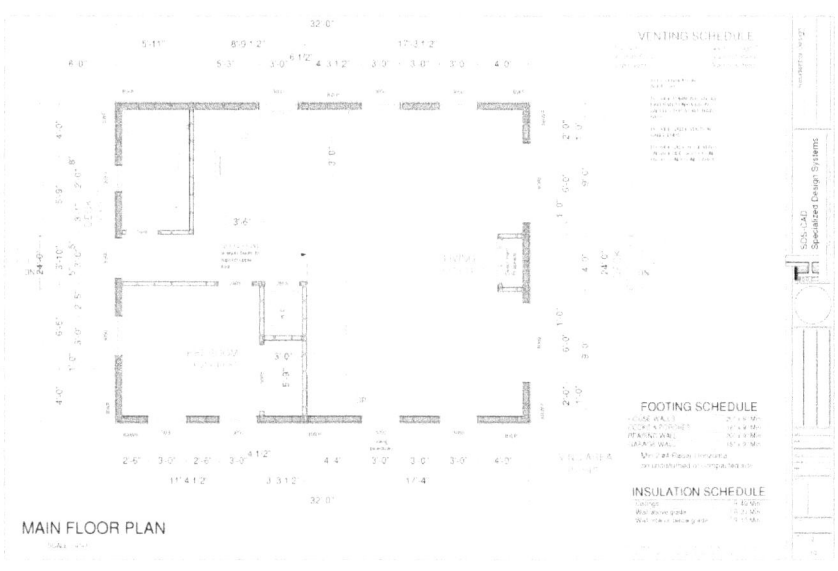

VENTING SCHEDULE

FOOTING SCHEDULE

INSULATION SCHEDULE

MAIN FLOOR PLAN

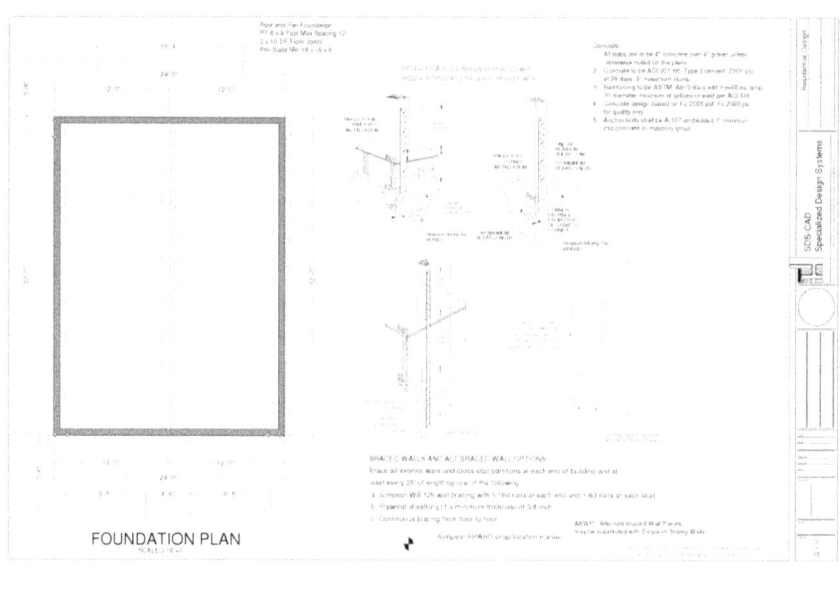

FOUNDATION PLAN

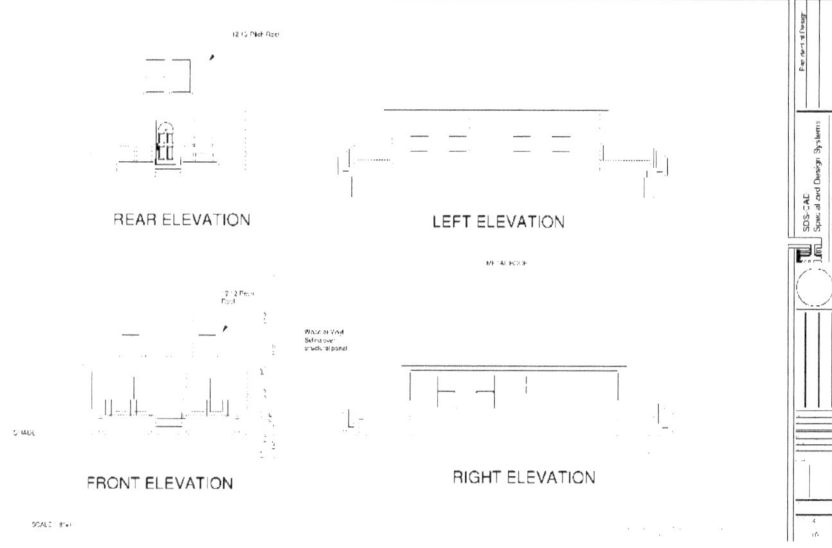

REAR ELEVATION

LEFT ELEVATION

FRONT ELEVATION

RIGHT ELEVATION

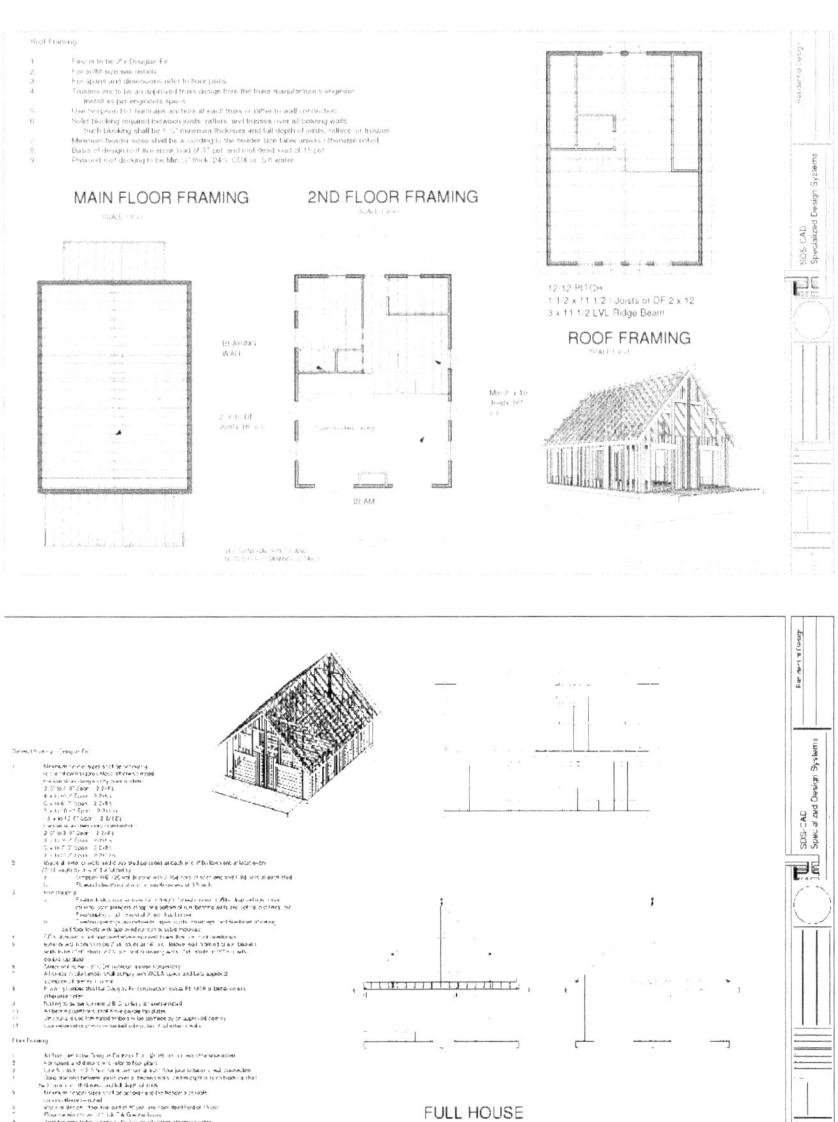

### MAIN FLOOR FRAMING

### 2ND FLOOR FRAMING

### ROOF FRAMING

### FULL HOUSE
### FRAMING SECTION

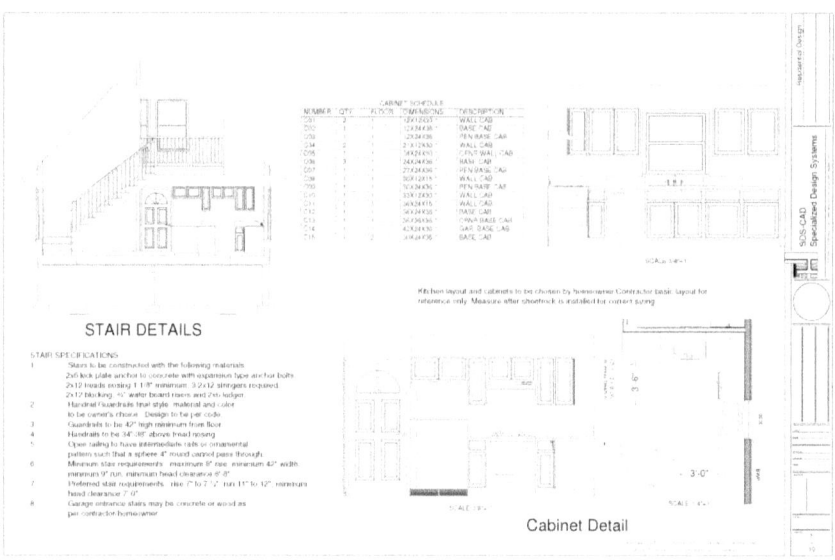

## STAIR DETAILS

STAIR SPECIFICATIONS
1. Stairs to be constructed with the following materials
   2x6 kick plate anchor to concrete with expansion type anchor bolts
   2x12 treads resting 1 1/8" minimum, 3 2x12 stringers required
   2x12 blocking, ¾" wafer board risers and 2x6 ledger
2. Handrail/Guardrails tread style, material and color
   to be owner's choice. Design to be per code.
3. Guardrails to be 42" high minimum from floor
4. Handrails to be 34"-38" above tread nosing
5. Open railing to have intermediate rails or ornamental
   pattern such that a sphere 4" round cannot pass through
6. Minimum stair requirements: maximum 8" rise, minimum 42" width,
   minimum 9" run, minimum head clearance 6'-8"
7. Preferred stair requirements: rise 7" to 7 ½" run 11" to 12", minimum
   head clearance 7'-0"
8. Garage entrance stairs may be concrete or wood as
   per contractor-homeowner.

Kitchen layout and cabinets to be chosen by homeowner/Contractor basic layout for reference only. Measure after sheetrock is installed for correct sizing.

### Cabinet Detail

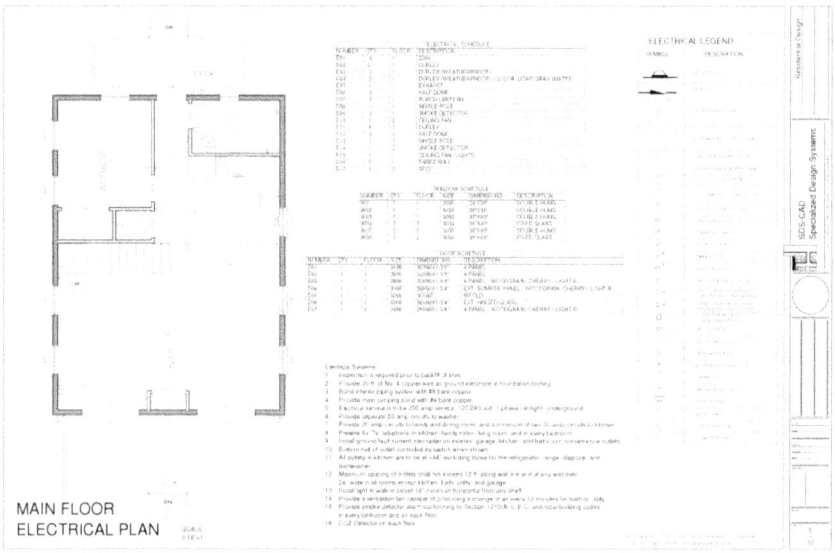

## MAIN FLOOR ELECTRICAL PLAN

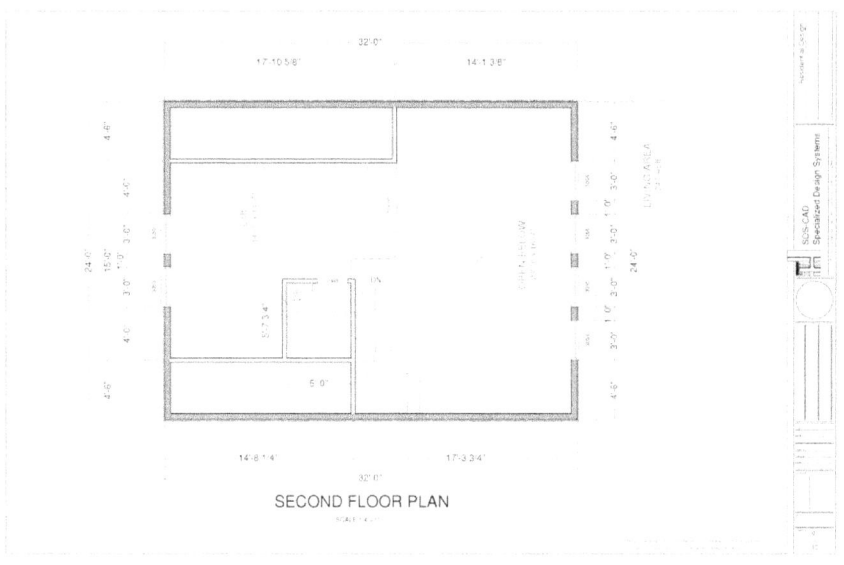

SECOND FLOOR PLAN

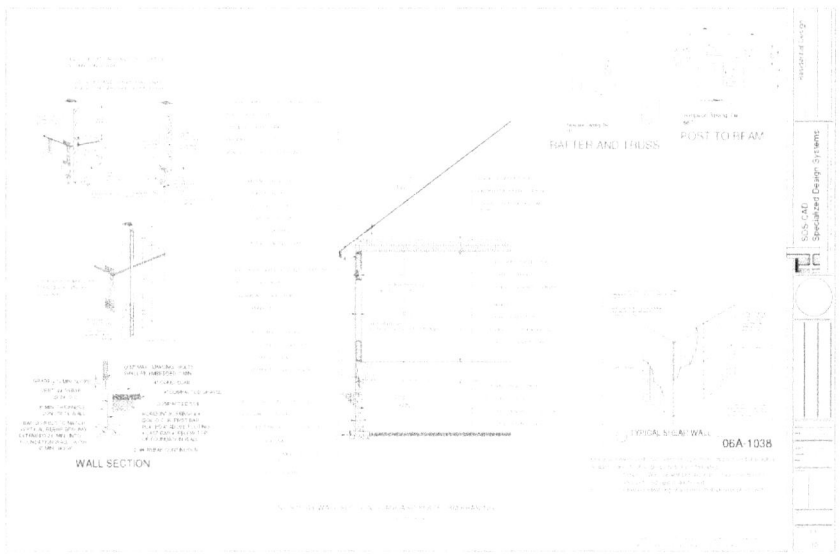

WALL SECTION

# 20 x 30 Small House Plans

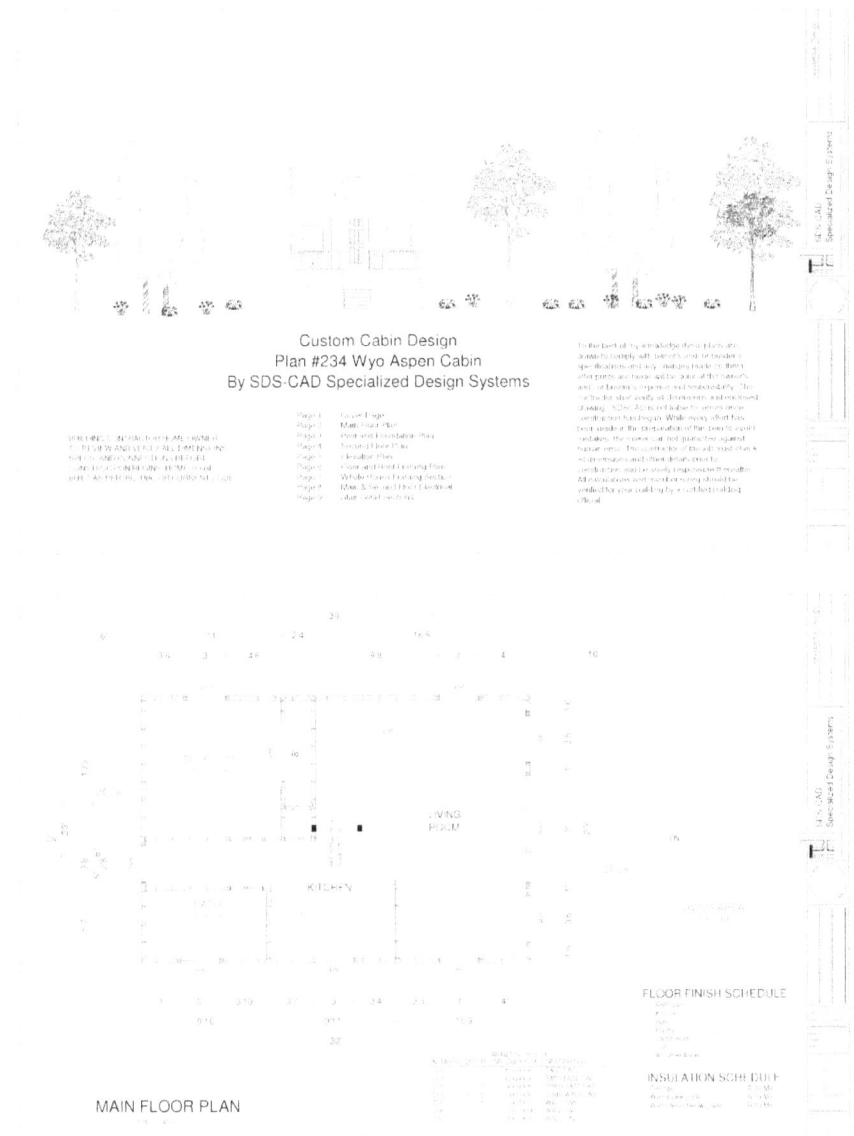

Custom Cabin Design
Plan #234 Wyo Aspen Cabin
By SDS-CAD Specialized Design Systems

MAIN FLOOR PLAN

FOUNDATION PLAN

SECOND FLOOR PLAN

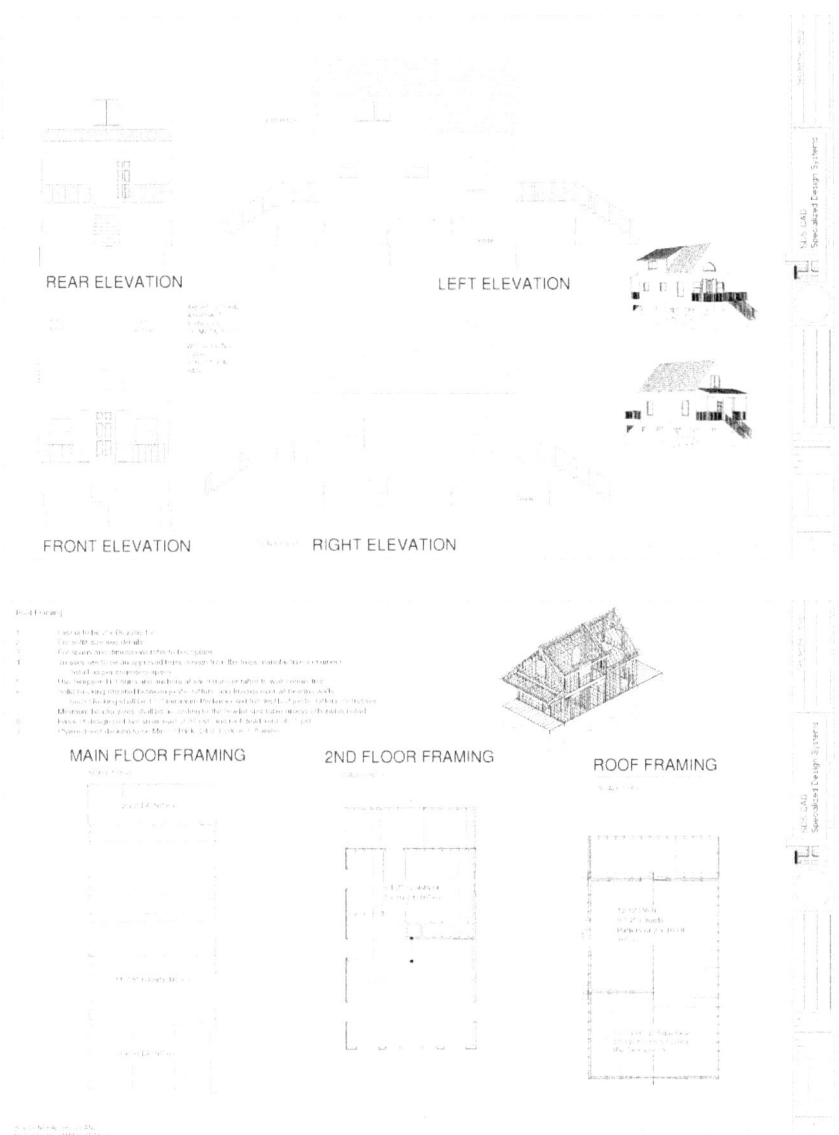

REAR ELEVATION

LEFT ELEVATION

FRONT ELEVATION

RIGHT ELEVATION

MAIN FLOOR FRAMING

2ND FLOOR FRAMING

ROOF FRAMING

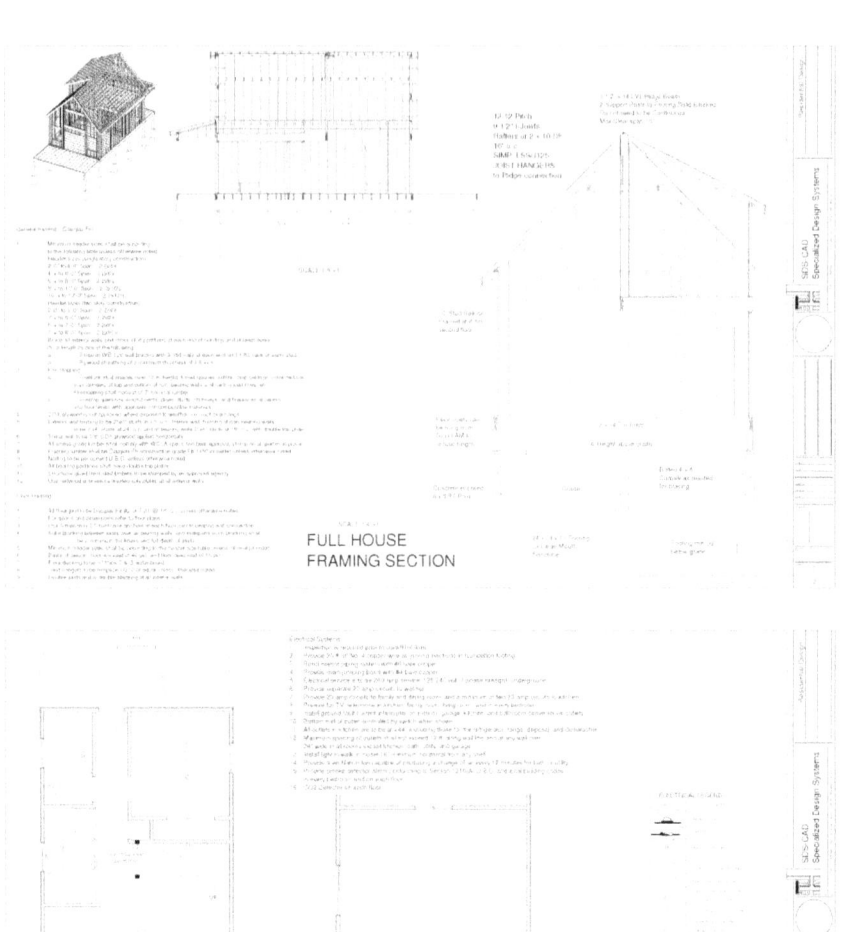

**FULL HOUSE
FRAMING SECTION**

**MAIN FLOOR
ELECTRICAL PLAN**

**SECOND FLOOR
ELECTRICAL PLAN**

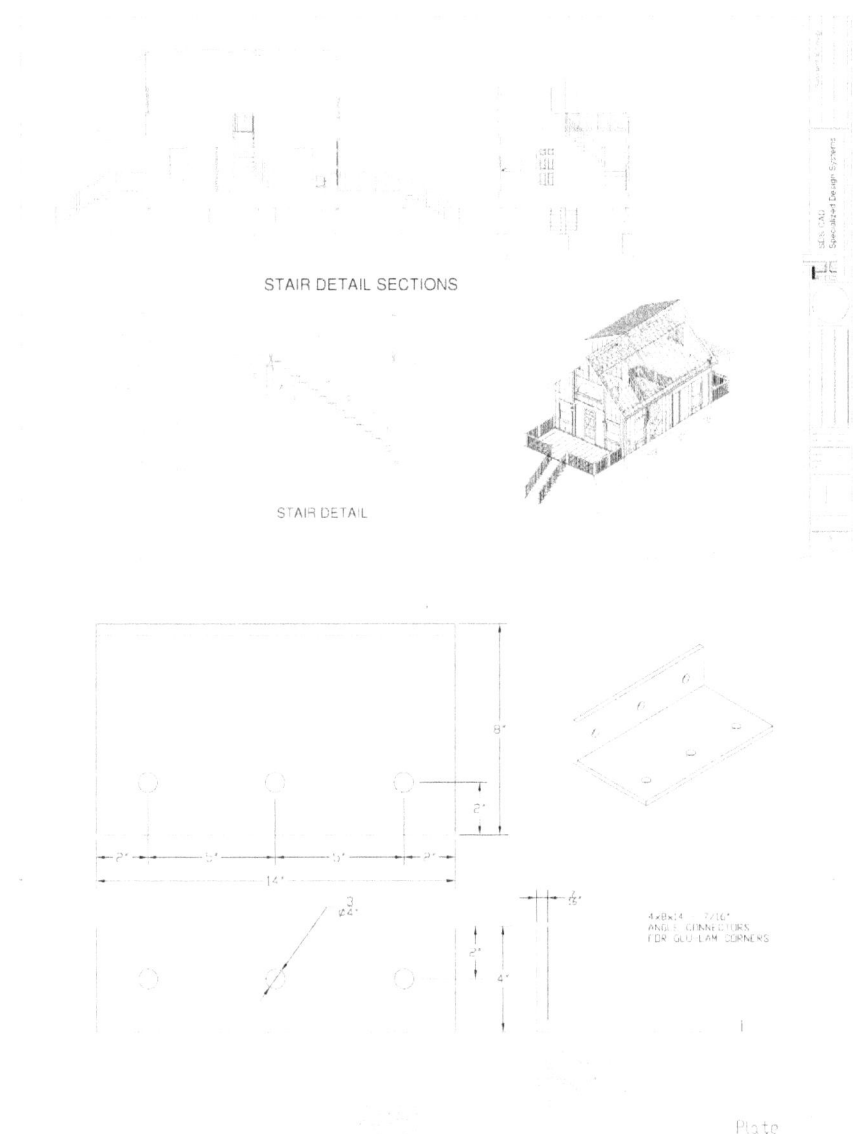

STAIR DETAIL SECTIONS

STAIR DETAIL

# 16 x 20 Bunkhouse Plans

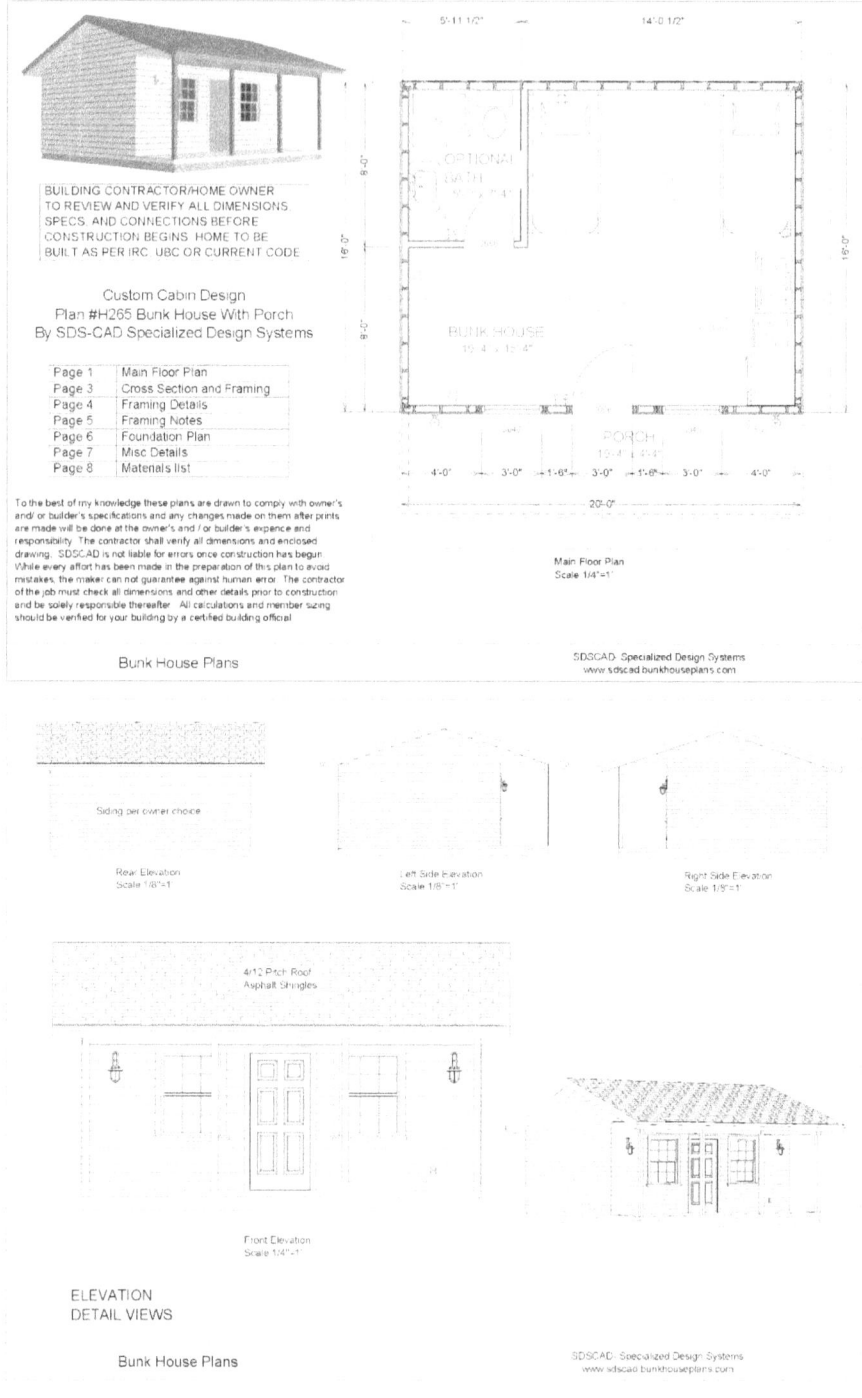

BUILDING CONTRACTOR/HOME OWNER
TO REVIEW AND VERIFY ALL DIMENSIONS,
SPECS. AND CONNECTIONS BEFORE
CONSTRUCTION BEGINS. HOME TO BE
BUILT AS PER IRC, UBC OR CURRENT CODE

Custom Cabin Design
Plan #H265 Bunk House With Porch
By SDS-CAD Specialized Design Systems

| Page 1 | Main Floor Plan |
| Page 3 | Cross Section and Framing |
| Page 4 | Framing Details |
| Page 5 | Framing Notes |
| Page 6 | Foundation Plan |
| Page 7 | Misc Details |
| Page 8 | Materials list |

To the best of my knowledge these plans are drawn to comply with owner's
and/ or builder's specifications and any changes made on them after prints
are made will be done at the owner's and / or builder's expence and
responsibility. The contractor shall verify all dimensions and enclosed
drawing. SDSCAD is not liable for errors once construction has begun.
While every effort has been made in the preparation of this plan to avoid
mistakes, the maker can not guarantee against human error. The contractor
of the job must check all dimensions and other details prior to construction
and be solely responsible thereafter. All calculations and member sizing
should be verified for your building by a certified building official.

Bunk House Plans

BUNK HOUSE
19-4 x 12-4

PORCH
19-4 x 4-8

Main Floor Plan
Scale 1/4"=1'

SDSCAD- Specialized Design Systems
www.sdscad.bunkhouseplans.com

Rear Elevation
Scale 1/8"=1'

Siding per owner choice

Left Side Elevation
Scale 1/8"=1'

Right Side Elevation
Scale 1/5"=1'

4/12 Pitch Roof
Asphalt Shingles

Front Elevation
Scale 1/4"=1'

ELEVATION
DETAIL VIEWS

Bunk House Plans

SDSCAD- Specialized Design Systems
www.sdscad.bunkhouseplans.com

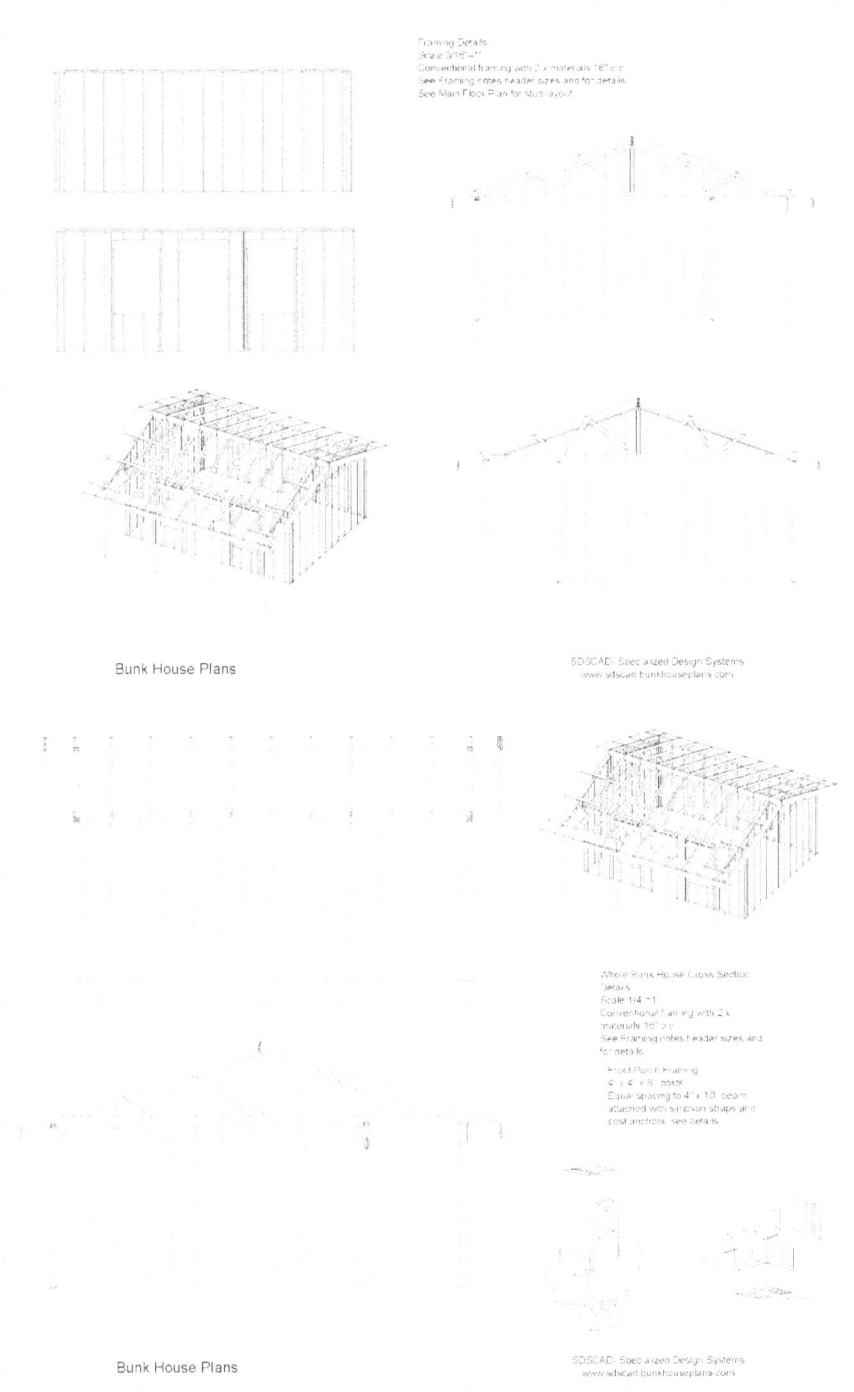

Framing Details
Scale 3/16"=1"
Conventional framing with 2 x materials 16" o.c.
See Framing notes header sizes and for details.
See Main Floor Plan for stud layout

Bunk House Plans

SDSCAD: Specialized Design Systems
www.sdscad-bunkhouseplans.com

Whole Bunk House Cross Section
Details
Scale 1/4"=1'
Conventional framing with 2 x
materials 16" o.c.
See Framing notes header sizes and
for details.

Front Porch Framing
4" x 4" x 8' posts
Equal spacing to 4" x 10' beam
attached with simpson straps and
post anchors, see details

Bunk House Plans

SDSCAD: Specialized Design Systems
www.sdscad-bunkhouseplans.com

General Framing (Douglas Fir)

1. Minimum header sizes shall be according to the following table unless otherwise noted.
   Header sizes (single story construction)
   2'-0" to 4'-0" Span   2-2x4's
   4' to 6'-0" Span   2-2x6's
   6' to 8'-0" Span   2-2x8's
   8' to 10'-0" Span   2-2x10's
   10' to 12'-0" Span   2-2x12's
2. Brace all exterior walls and cross-stud partitions at each end of building and at least every 25' of length by one of the following.
   a.    Simpson WB 126 wall bracing with 3-16d nails at each end and 1-8d nails at each stud
   b.    Plywood sheathing of a minimum thickness of 3/8 inch
3. Fire stopping
   a.    Firebrick stud spaces over 10' in height, furred spaces, soffits, drop ceilings, cove ceilings, stair stringers at top and bottom of run, bearing walls and ceiling joist lines, etc. Firestopping shall consist of 2" nominal lumber.
   b.    Firestop openings around vents, pipes, ducts, chimneys, and fireplaces at ceiling and floor levels with approved noncombustible materials.
4. CDX plywood is not approved where exposed to weather, i.e. roof overhangs.
5. Exterior wall framing to be 2'x4" studs Min at 16" o.c. Interior wall framing at non bearing walls to be 2'x4" studs at 24' o.c. and at bearing walls 2'x4" studs at 16" o.c. with double top plate.
6. Shear wall to be 3/8" CDX plywood applied horizontally.
7. All stress grade lumber shall comply with WCLA specs and bear approval stamp on all pieces in place.
8. Framing lumber shall be Douglas Fir construction grade Fb 1450 or better unless otherwise noted.
9. Nailing to be per current U.B.C. unless otherwise noted.
10. All bearing partitions shall have double top plates.
11. Structural glued laminated timbers to be stamped by an approved agency.
12. Use redwood or pressure treated sole plates at all exterior walls.

Roof Framing

1. Fascia to be 2'x4" Douglas Fir
2. 12" soffit size
3. For spans and dimensions refer to floor plans or default to 16" o.c. framing 24' o.c. trusses.
4. Trusses are to be an approved truss design from the truss manufacture's engineer.
5. Use Simpson H-1 hurricane anchors at each truss or rafter to wall connection.
6. Solid blocking required between joists, rafters, and trusses over all bearing walls. Such blocking shall be 1 ½" minimum thickness and full depth of joists, rafters, or trusses.
7. Minimum header sizes shall be according to the header size table unless otherwise noted.
8. Basis of design roof live/snow load of 37 psf and roof dead load of 15 psf.
9. Plywood roof decking to be ½" thick, 24/0 CDX of 7/16 wafer.

Bunk House Plans

SDSCAD: Specialized Design Systems
www.sdscad.bunkhouseplans.com

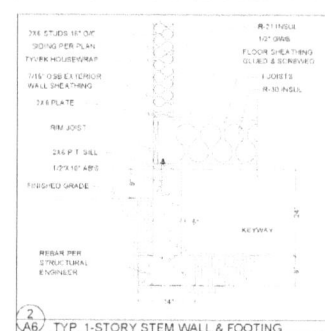

2 / A6    TYP. 1-STORY STEM WALL & FOOTING

Optional Foundation System

Monolithic Concrete Slab Plan
Scale 1/4 =1'

Bunk House Plans

SDSCAD: Specialized Design Systems
www.sdscad.bunkhouseplans.com

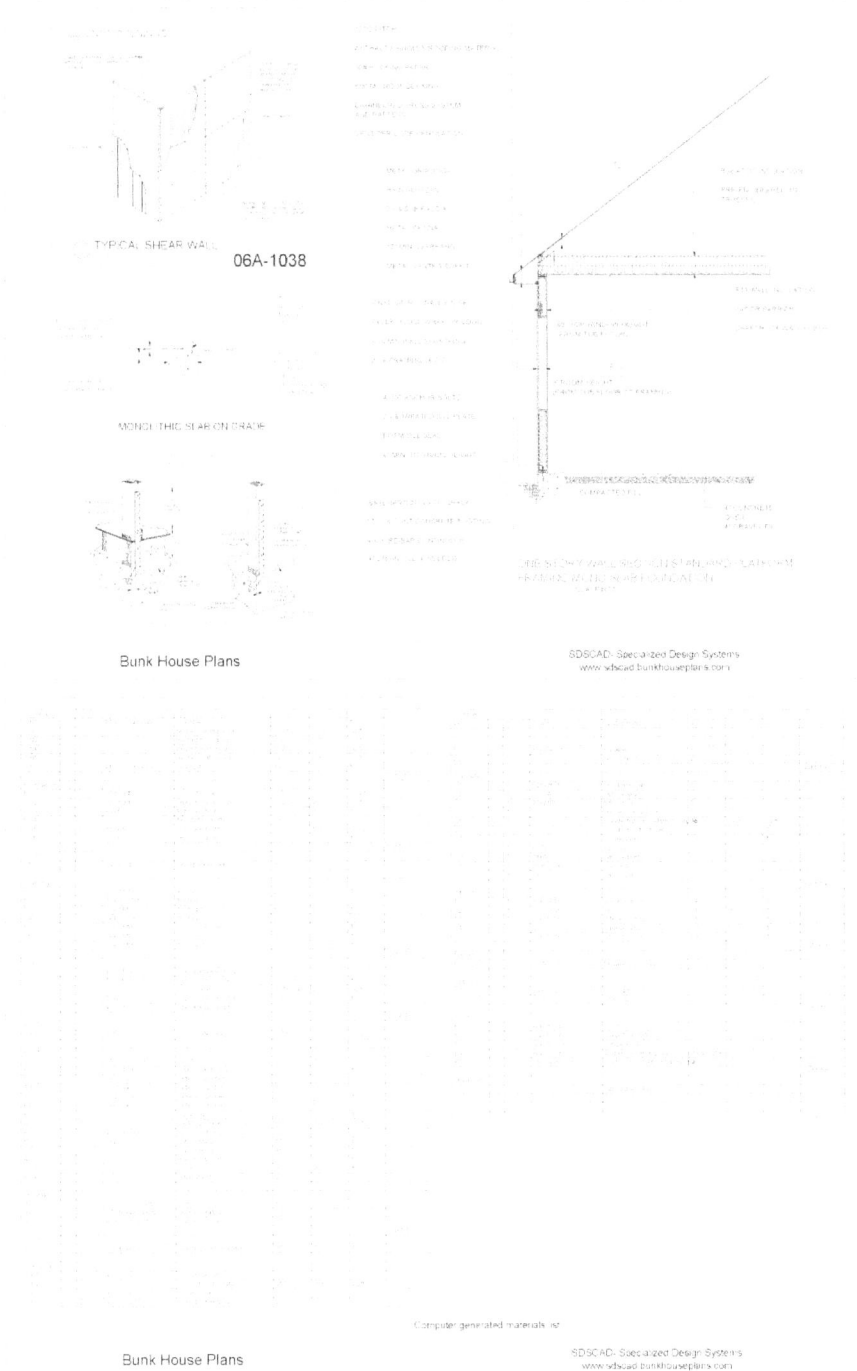

TYPICAL SHEAR WALL

06A-1038

MONOLITHIC SLAB ON GRADE

Bunk House Plans

SDSCAD- Specialized Design Systems
www.sdscad.bunkhouseplans.com

Computer generated materials list

Bunk House Plans

SDSCAD- Specialized Design Systems
www.sdscad.bunkhouseplans.com

# Building Process Pictures for our bunkhouse

The Following are the pictures of our family building a small 16 x 16 bunkhouse up in the mountains at our family cabin over a two days.

We hand built our own trusses on the driveway so they would be flat before heading to the mountains to build the small bunkhouse

We sided the house using rough sawn cedar board and baton siding.

Building bunks and putting plywood up on the inside walls

# Author Bio

## Colvin Tonya Nyakundi

Colvin Tonya Nyakundi is a freelance writer and co-author of 'How to Build a Simple Wooden House' Apart from that book, he has a portfolio of several other publications accumulated in the more than two years that he has been freelancing through www.odesk.com.

In addition to his interest in construction and real estate publications he has authored several personal relationships, lifestyle and travel and holiday guide publications. Other books that he has co-authored include 'How to Survive in the Woods', 'How to Start Making Money Online', 'How to Survive in a Desert', 'How to Improve Your Communication Skills,' 'Construction Guide for New Investors in Real Estate,' 'How to Make Your Backyard a Magnificent Venue for Hosting Events', 'How to Identify the Perfect Holiday Destination', "How Your Favorite Meal Could be Killing You Slowly" and 'How to Prepare and Survive in a Foreign Country.' You can get in touch with him through his official Facebook account, tonyanc@facebook.com.

Check out some of the other JD-Biz Publishing books

Gardening Series on Amazon

# Health Learning Series

Our books are available at

1. Amazon.com

2. Barnes and Noble

3. Itunes

4. Kobo

5. Smashwords

6. Google Play Books

## **Publisher**

JD-Biz Corp

P O Box 374

Mendon, Utah 84325

http://www.jd-biz.com/

www.ingramcontent.com/pod-product-compliance
Lightning Source LLC
Chambersburg PA
CBHW070831180526
45168CB00002B/799